Ron Mueller

Stress Free™ **Changeover Solutions**
Throughput Optimization

By: Ron Mueller

Around the World Publishing LLC
4914 Cooper Road Suite 144
Cincinnati, Ohio 45242-9998

ISBN 13: 978-1-68223-246-0
ISBN 10: 1-68223-246-8

Distributed by Ingram
Cover Picture By: Lisa F. Young, Dreamstime.com
Cover Design by: Ron Mueller

Ron Mueller

TECHNICAL EDITOR:

Gordon Miller P. E.

vii

DEDICATION

*To all those who want to make it
fast,
easy to do
and
error free.*

Table of Content

Ron Mueller

xi

ACKNOWLEDGMENTS

The people put in a life time of hard work.

Making Changeover, easier and faster to do makes

this work a little easier.

Everyone doing the work deserves this break.

Ron Mueller

Introduction

The concept of rapid changeover is very well demonstrated in car racing's pit row rapid tire changes and refueling. The coordination, teamwork, equipment and wheel preparation with the car is exceptional. The gravity feed gas resupply is faster than can be achieved with a pressurized system. These processes are done by very talented people, who have practiced to the point of perfection. They minimize the pit time and maximize the run time and thus directly affect the outcome of the race.

This concept can be applied along the entire supply chain; at the customer, at the truck loading area, at the finished product warehouse, at the production line, and for raw material transport and receiving. It can also be applied in many other repetitive work processes.

The race is about meeting the customer product demand as fast and as cost effective as possible. The time and the quantity of raw materials and finished product that is either in transit or in storage along the supply chain affects the profit margin for the product and if possible must be reduced.

The supply chain usually has multiple material flows and multiple finished products being produced. The supply chain in fact is made up of multiple supply chains that flow the physical material to product(s) transformation system(s). This parallel, multi-product supply system has a direct impact on the profit for the company.

In manufacturing, changeover is the process of converting a line or machine from running one product to another. Changeover times vary based on the design of the production equipment. The goal is to make it as short as economically possible.

Reducing changeover time creates more productive (value-added) time for production. Additionally reducing changeover time allows reduction of production batch sizes, work-in-process (WIP), and inventory.

Changeover is the amount of time taken to change a piece of equipment from producing the last good piece of a production run to the first good piece of the next production run.

Changeover focuses on:

- Improving the timeliness of response to customer orders.
- Quality on the first next product.
- Improving the flow of the supply chain.
- Reducing inventory allowing the recovery of operating cash and manufacturing space.
- Increase production flexibility
- Increasing Throughput
- Eliminating setup adjustments

The required or target changeover time is determined by:

- What it takes to satisfy the customer.
- The available time and number of products to be produced on the same line.
- The cost of the changeover
- By the product family value stream capability.

Financially it comes down to the choice of spending capital to gain more capacity, or storing more finished product to be able to ship to order or to optimizing the current capital investment by producing multiple products on one production line.

This generates questions such as;

- How much inventory is required for each product?
- How many changeovers are required to meet the customer demand?
- Would investing in more capacity be a better choice?
- Do I need to move inventory from DC to DC?

These are questions that determine the required changeover time. These are questions that leaders will answer at they set the changeover time targets.

The goal is to achieve a low variability, cost effective flow product to the customer.

What Flows?

"ITEMS" flow through a value stream

- In manufacturing, materials and finished product are items.
- In design & development, designs are the items.
- In service, external customer needs are the items.
- In administration., internal customer needs are the items.

Changeover Process

Changeover is focused on improving the supply chain flow.

Stress Free^{TM} Changeover Solutions (COS) is presented in Six steps.

Change Over Process Process

Step 1: Business Need
- Opportunity Dimensions
- Changeover Focus Scope
- Preparing Changeover Improvement Participants

Step 2: Understand the Current Situation
- External changeover preparation
- Shutdown preparation
- Internal changeover
- Startup
- External Completion

Step 3: Make Improvement
- Separate Internal from External
- Stream line the internal
- Stream line the external

Step 4: Validate the Improvement
- Organizational Improvement
- Equipment Improvement

Step 5: Practice the Changeover

Step 6: Integrate into daily management system.

Chapter 1: Business Need

Step 1: Prepare for Changeover Improvement Leadership

Changeover improvement must be tied to a specific business need. The drivers for improving the time it takes to move from the production of one product to the production of the next product are:

1. The need to produce a greater number of products on the same line.
    The ability to produce multiple products on the same line helps reduces the capital that must be spent to build multiple lines.

2. The desire to free cash trapped in inventory.
    The ability to make the product on multiple smaller volume production runs allows the inventory to be kept at a lower level.

3. The desire for better on time delivery.
    The ability to respond rapidly to an order enhances the organizations ability to meet variations or sudden requests that take the inventory level to zero.

Need Statement Steps

Establish Changeover goal

What does the business need

Be aggressive – an improvement effort should yield 80% improvement

What do the operators need? The changeover improvement should make it easier for them.

Value of the Changeover

The value of improving the changeover is determined by several factors;

1. The dollar value of the finished product inventory.

2. The value of in-process materials

3. The value of the raw materials inventory

   These three are also referred to as trapped cash. This is money that could be freed up and potentially used to invest in growing the business or paying for improvement efforts.

4. The impact on customer service

    The ability to quickly respond to changing customer demand with no financial impact enhances the business performance.

5. The impact in the ability to be flexible and responsive to change.

    This is mentioned again for emphasis. A business that has the ability to respond with no loss is able to partner with their customers and improve the supply chain for all participants.

Business Linkage

This changeover concept is utilized in the product production area to facilitate the changeover from one product to the equipment or process set up for a second and different product. This changeover capability allows the reduction of finished product inventory and leverages the existing capital tied up in the equipment.

It is important to understand that changeover improvement comes from a variety of areas.

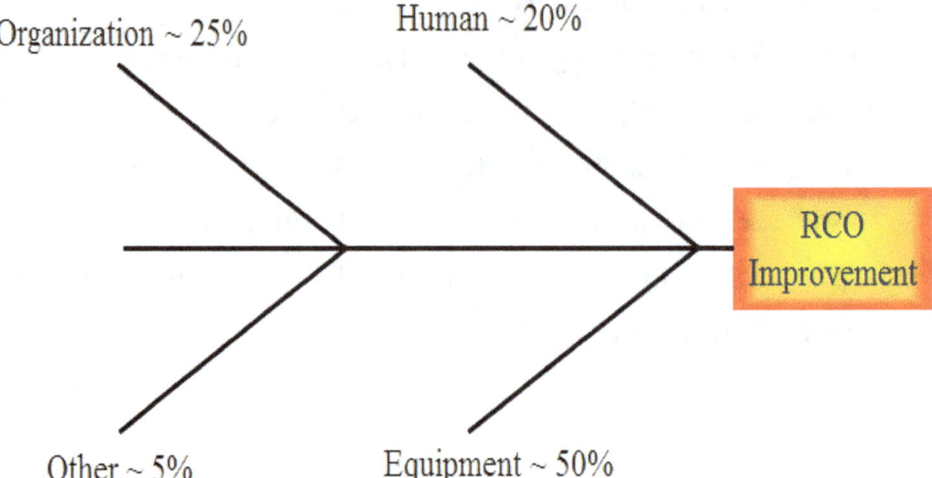

Organization Connection

Leaders define and set the behavior and culture of the organization. The organization design determines how it responds to the required work and to the variations that impact the work. Leaders must understand their supply chain as well as their business processes.

A leader's number one focus is to take the actions that ensure their customer's needs are met and exceeded. This customer focus drives the design of the entire organization.

The people involved in the work to satisfy the customer need to be enabled by a developmental, skill growth culture. The design of such a culture is detailed in *"Stress Free™ Organization and Leadership Solutions*. This book will focus on optimizing the work of changing over from a specific series of work and equipment processes to another in as short a time as possible.

Human Connection

Humans and equipment do work. Equipment is used in the broad sense that includes computers and other data processors as well as the equipment that transforms raw materials into finished products.

Humans grow in capability over time. Equipment and processes must be developed by the humans. The growth of the human is determined by the leaders of the organization as well as the human's own drive and interest. The organization that stalls and fails, often does so because of leadership's failure to cultivate and grow the human capability.

This book focuses on developing both the human capability as well as simplifying the equipment and process to improve the speed of the changeover.

Equipment Connection

Changeover time is determined by equipment design and installation lay out. Reducing change over time requires that the human to equipment interaction be improved by making each task easier to do. Changeover time is reduced, when the work required to do it is reduced.

System Changeover

The changeover cycle is defined as the time it takes to go from "going to going". The changeover time begins when the production stop button is pressed. The changeover ends when the product is being produced at the target production rate.

The system changeover may be going on in many places along the supply chain. These changeovers need to be synchronized to insure they are not affecting each other.

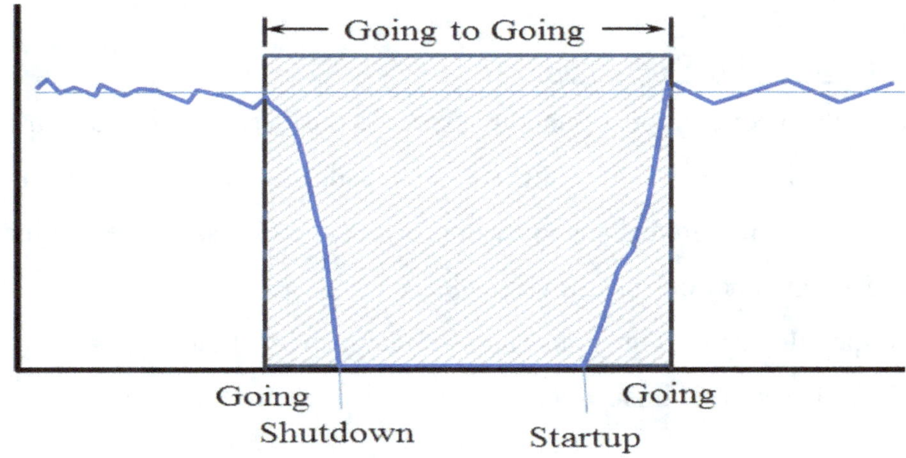

The going to going time is the focus for the improvement effort. Ideally it would be close to a zero gap in the production process.

System Layout

Changeover (CO) capability improvement is important when implementing statistical inventory replenishment. CO allows for the reduction of production pitch or production quantity orders. This in turn facilitates the reduction of inventory and improves the ability to serve the customer on time with at least a 99.7% customer delivery.

The concepts in *Stress Free*[TM] *Changeover Solutions* are successful because of the second by second granularity of timing and analyzing the process of changing the equipment over. This increased granularity from hours and minutes to seconds provides new insights to improvement opportunities.

This approach builds on concepts presented in Stress Free[TM] *Work Process Solutions.*

In Changeover (RO) improvement results in;
1. Reduced work effort.
2. The ability to reduce inventory
3. The ability to increase throughput.
4. Increase of on time product delivery

This provides a reward to everyone doing the work and a significant gain for the business.

Improvement Value Evaluation

Tying changeover improvement to a specific business need or goal ensures the effort will be valued. This value should be judged on the basis of a business investment. The required changeover improvement choices can then be judged the same as any other business investment.

The specific business linkage drives the changeover strategy. The cost of changeover helps to set this strategy.

Value Measurement

Cost

Changeover Cost = [(Average Changeover time) x (Line Speed)(y products/min)] x [Cost per y product ($/product)]

The changeover cost coupled with the number of required changeovers and the inventory carrying costs can then be utilized to determine the value of making any improvement.

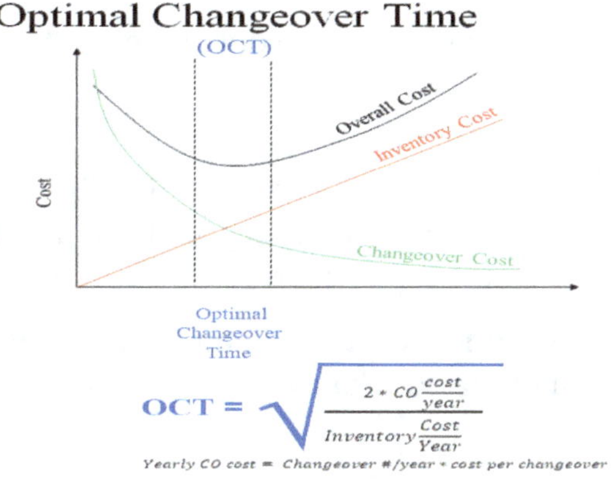

Being able to reduce the changeover time and therefore the cost of changeover moves the changeover curve on this diagram.

Changeover Preparation

The value to the business has been determined. Leadership is clear that an improvement to the changeover process will be important. An improvement team has been selected.

The team should be comprised of those doing the changeover and the people that will participate in improving the changeover. The changeover team should be paired with a person that will be observing and documenting the changeover action.

The team will need training that will include the changeover process and how success will be measured. This measure will be on value. Value to the person and value to the business.

Preparation insures the success of the changeover improvement.

The Preparations Focus on:

People

Each person on the team will have key roles that they must fulfill.

Safety Leader: must ensure all participants know and follow the safety requirements.

Change agent: Also known as the "devil's advocate" for constantly asking why or why not?

Support agent: This is the person who knows how to get stuff when needed.

Timing agent: Anyone in the observation role.

Documentation agent: the person willing to document each step of the process.

Often one person will fill several roles or people will partner up.

<u>Role Clarification</u>: Everyone needs some specific training. This is provided by the workshop trainer.

<u>Role Walk Through</u>: The trainer will also do an on the floor walk through of standardized Change Over Tasks

<u>Role Floor Check</u>

Proper parts in place
Needed Tools in place
Extraneous materials out of the way

Materials

Quantity – minimum required

Floor Location

Tools

Needed tools

Organized

Placed in the proper location

Change Parts

- Right parts
- Right condition
- Right Place
- Easy to verify

Walk-Through

Review Changeover each individual at their changeover location

It is important to get to the floor and watch the actual changeover in progress. The line team each have an observation partner. The line team member and the partner do a simulation of what the changeover entails. Production continues undisturbed.

External Changeover Preparation

This is the first part of the walk through. Each changeover location may have slightly different preparation. The cleanup procedure, the change part staging is described.

Shutdown preparation

The details of the shutdown process is described. Safety procedures are a key part of this preparation. Control system, mechanical, electrical, hydraulic and pneumatic lock out procedures are reviewed.

Internal Changeover

A step by step description of the changeover procedure is given by the changeover person. The changeover partners for the improvement process ask any clarifying questions.

The description is done with the production system running.

The end of a production line changeover is a point in time where extraordinary safety consideration needs to be observed.

Startup

Startup must be designed in a mistake proof fashion. Every individual on the changeover team has their personal lockout. Each individual must have a foolproof way to let the line leader know that it is safe to startup.

External Completion

The clean up around the production line and of the change parts is an important step but may not be as time dependent as the internal changeover duration.

Make final preparation

Initial improvements may already have been identified. These initial ideas may be tried and evaluated. These initial ideas need to be documented and the modified procedures talked through.

Limitations

Equipment

The intent is to experience the first changeover trial with no equipment change. Subsequent changeovers may require minor changes but they will be those changes that can be made immediately such as;

- Change parts location
- Tool location
- Safety cover bolt number reduction
- Quick disconnect application.

Any changes must be reviewed for safety and approved as safe to do.

Material

Material placement and movement out of the way are often important to consider. Safely handling material movement away from the change area and then returning prior to startup should be considered where the space is tight.

System

Any control system changes that are required to run the next product are a part of the changeover process. Often these changes must be documented as a way to verify the changeover.

Legal/Cultural

Some businesses have legal changeover requirements. These must be specifically adhered to. However even the legal requirements should be examined to ensure they were properly interpreted.

There are often environmental or work process limitations. The changeover teams are often limited by the number of available people and the workspace around the production line.

Additionally, there may be a variety of legal requirements that impact the changeover process or speed.

Define the required Changeover time.

Defining the required changeover time is key step in determining if Changeover improvement is warranted. This required changeover time can then serve as a lens by which to evaluate all transformation points along the supply chain. This ensures the improvement at one point along the supply chain can be properly supported and maintained.

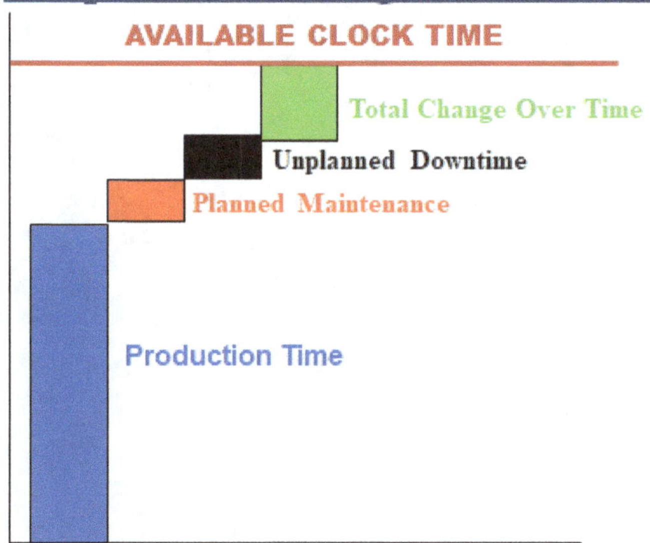

1. Determine Available Clock Time
    a. Begin with total Time
    b. Remove Lunches and Breaks
    Note: CO, PM and Unplanned Time is include in the available time.
2. Determine the time to produce the customer required product (Production Time)
3. Subtract Production Time from the Available Clock Time.
4. Subtract Planned Maintenance from the Available Clock Time.
5. Subtract Unplanned Downtime from the Available Clock Time.
6. What Remains is the Total Change Over Time.
7. Divide the Total Change Over Time by the Target time for Changeovers.
    - This result is the **TOTAL NUMBER OF CHANGE OVERS THAT CAN BE SCHEDULED.**
8. Divide the Total Change Over Time by the number of changes required to run the scheduled production.
    - This result is the **Required Change Over Time.**

<u>Chapter 1 Tips and Tools:</u>

1. Use Case Study File / Workbook
2. Determine the changeover value
3. Determine the changeover success measure
4. Do a mental, verbal and on the floor walkthrough

Chapter 2: Current Situation

<u>Step 2 Current Situation</u>

Now it is time to do it. Each step will be studied and documented.

The following documents will all be used;

- Time Observation Sheet
- Travel Diagram
- Detailed Motion Chart
- Task Chart (alias Zig–Zag) chart
- Brainstorm Fishbone
- Change Over Observation

Each document will peel back the cover and expose the root cause solution.

Time Observation Sheet

This will provide detailed times for each movement the changeover requires. The external steps will be documented separately from the internal steps.

TIME OBSERVATION SHEET

Operator:

Date: Observer:

Tasks	Step	Steps	Observed Step Times □ sec	Lowest Repeat. time	Lowest Time	Highest Time	Range.	Travel Distance	Travel Time	Notes
1 Start		Log-out		0			0			
2		Lockout					0			
3		Electrical					0			
4		Mechanical					0			
5		Pnuematic					0			
6		Control System					0			
7		Remove Safety Guard					0			
8 Collator		Vacuum equipment					0			
9		Examine equipment					0			
10		begin change					0			
11							0			empty automatically
12		unbolt tray					0			
13		- remove tray and put					0			
14 Tampon chain		remove pusher					0			

Preparation of the Time Observation sheet

Have the changeover person describe every task and the steps of that task that they will be taking. This listing will be adjusted during actual observation if additional actions are necessary.

Travel Diagram

The travel diagram is used to understand the human movement at the high level. The travel diagram is very visual and allows for the immediate identification of improvements.

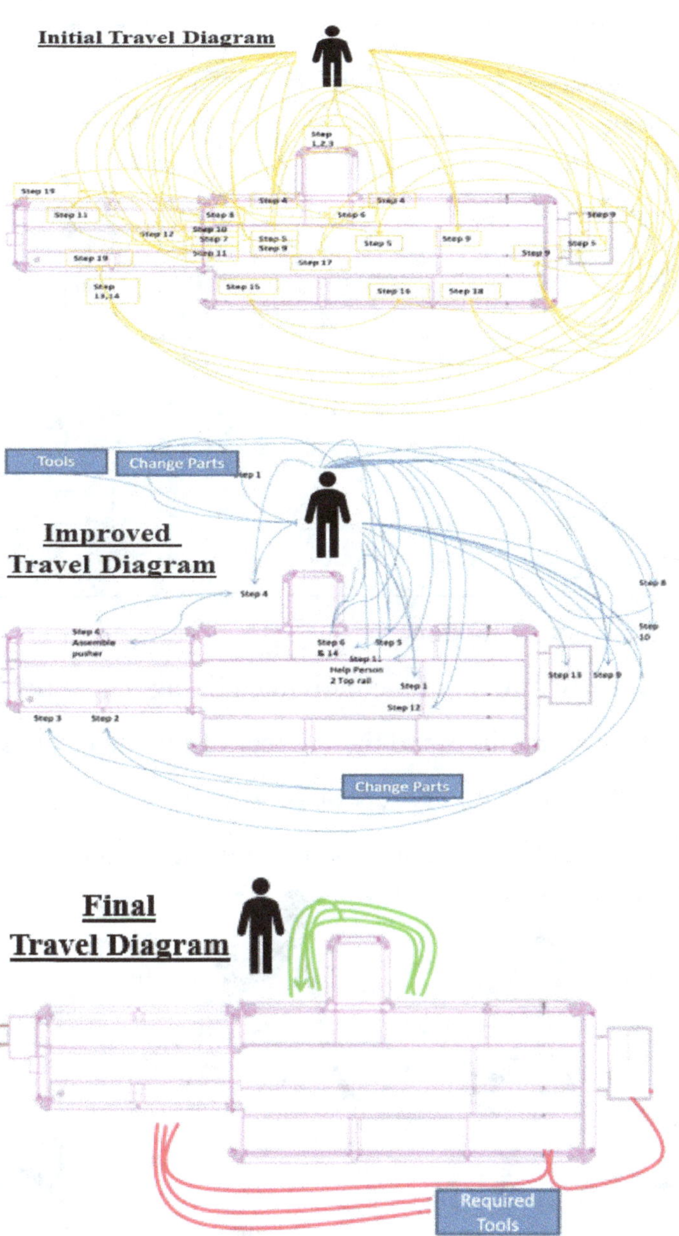

Detailed Motion Chart

The detailed motion diagram is the tool for the motion when the person is standing or sitting in one place doing the work.

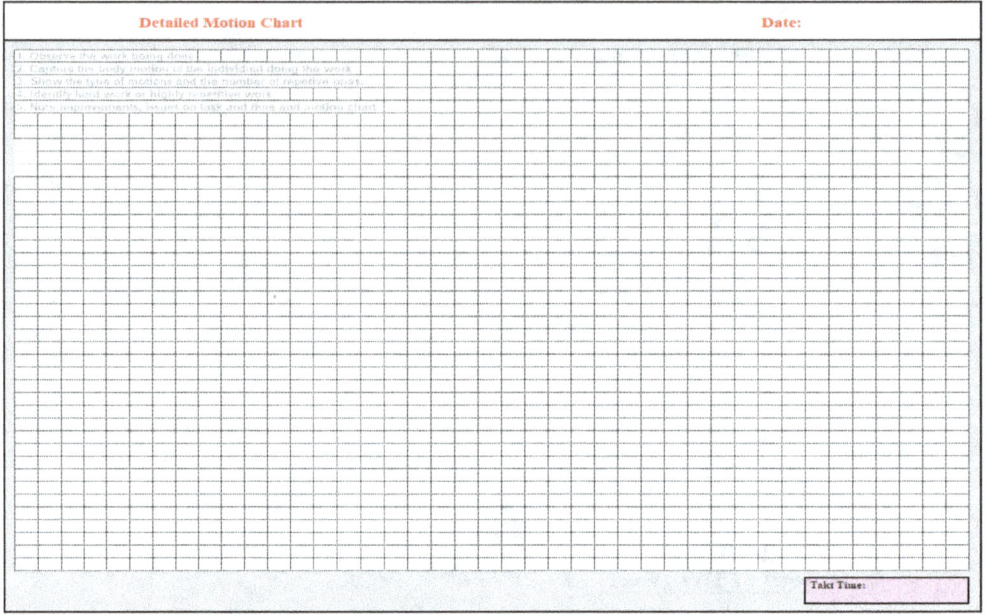

1. Observe the work being done

2. Capture the body motion of the individual doing the work

3. Show the type of motions and the number of repetitive ones.

4. Identify hard work or highly repetitive work

5. Note improvements, issues on task and time and motion chart

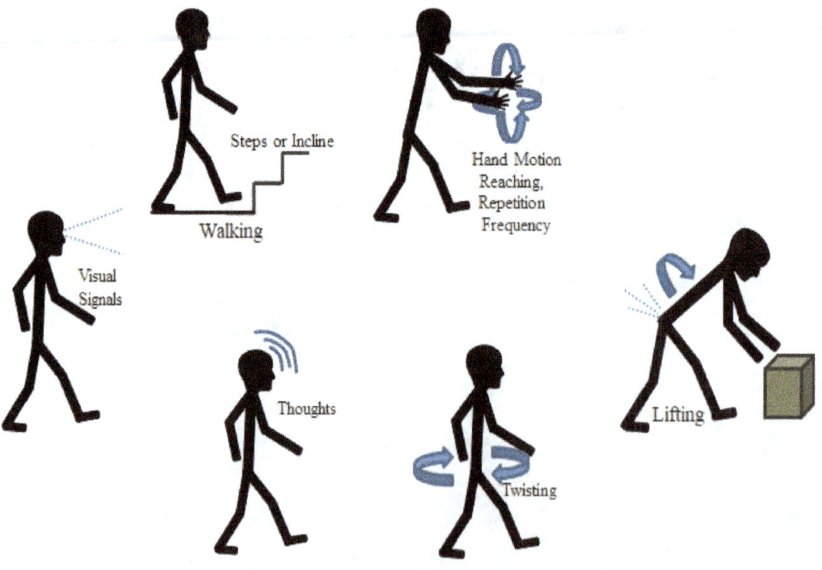

Task Chart (alias Zig–Zag) chart

This provides a visual of the value add and required steps. It is more often used in the office area or in a document management area.

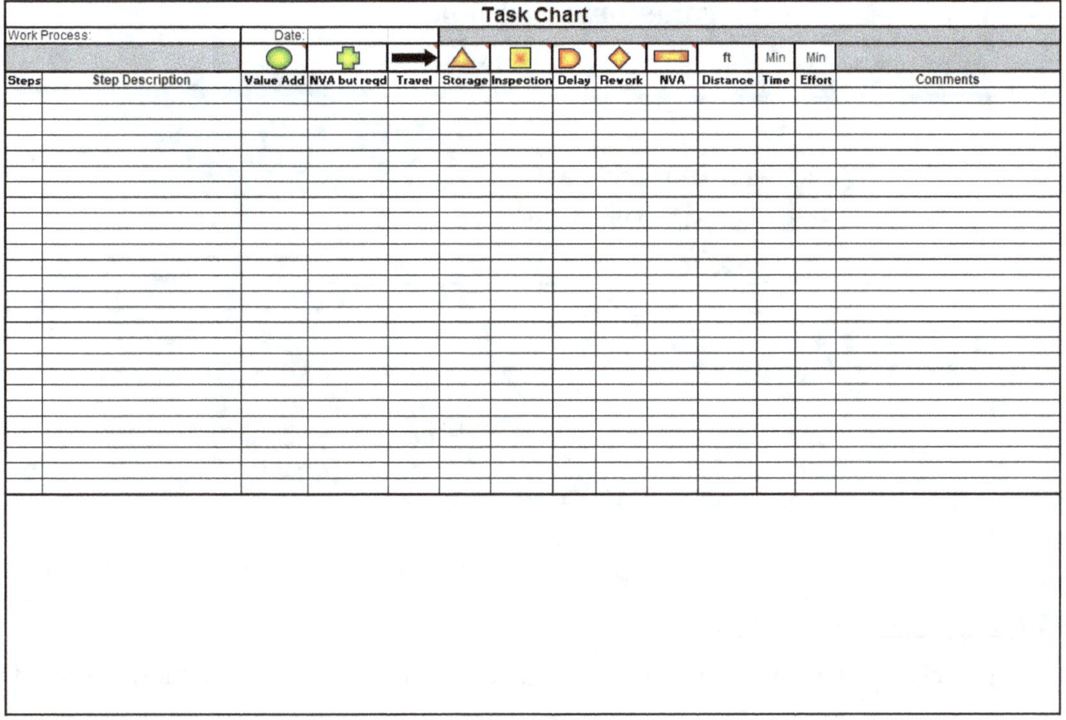

Brainstorm Fishbone

Changeover is one of the few places that I recommend brainstorming. The fishbone helps to organize the element of the brainstorm.

Brainstorm Fishbone

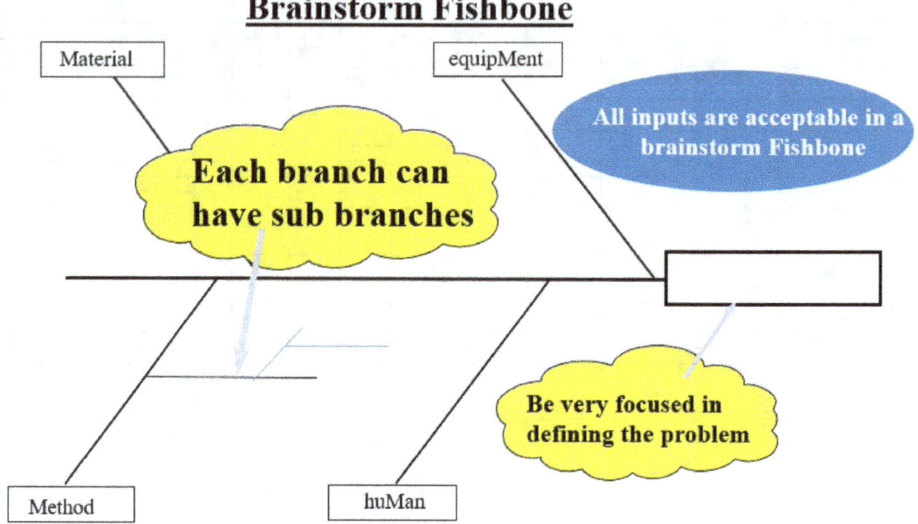

Change Over Observation

The changeover will be slowly executed and detailed documentation created or updated. This slow execution is a time when the process is examined for its ease of execution or any issues. It is a time that is rich with improvement discovery.

Use a separate detailed observation chart for each section.

If there are enough resources have one person timing each step in seconds, one person capturing the motion, one person immediately documenting. It is very common to have three persons teaming up to document the changeover.

- **External Preparation**
  This preparation will be done for the entire production line. Use the detailed observation chart.
- **Shutdown**
  The changeover improvement teams will synchronize their actions and shut down the line as one group. Safety procedure will be strictly followed and lockout will include all elements.
  Once the lockout is complete the internal changeover begins.

- **Internal Changeover**
  Each section of the line has a specific team to address the changeover. One team member will do the physical part of the changeover while the other documents the action.
  Once all teams have completed the changeover the line will be restarted by the operating team. All safety procedures are strictly adhered to
- **Startup**
  The all clear must be declared by each changeover improvement team. The line leader points to and acknowledges the clear signal from each improvement team.
- **External Completion**
  The changeover teams continue their improvement work until all the external work complete.
  This initial look may seem detailed. It is done this way in order to develop a

deeper understanding of the changeover processes. It is executed slowly and

should be studied thoroughly.

Chapter 2 Tips and Tools

1. Observe the Changeover

Chapter 3: Identify Improvement Opportunities

Step 3 Identify Improvement Opportunities

Create Changeover Timing Chart

Document the action of each changeover person (measure to seconds)

a) Time Observation Chart (example actions observed)

1. Walk to electrical lockout   ________ sec

2. Lock out electrical   ________ sec

3. Walk to pneumatic   ________ sec

4. Lock out pneumatic   ________ sec

5. Walk to operator side   ________ sec

6. Open guard.   ________ sec

7. Turn around and go to table   ________ sec

8. Select specific Allen wrench   ________ sec

9. Go back to filler   ________ sec

10. Open three locking screws   ________ sec

    Etc. usually between 50 to 100 steps

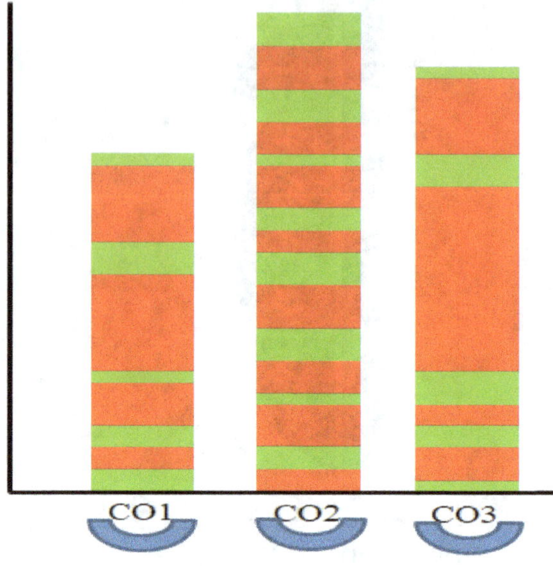

This timing chart would be the focus for the entire changeover. After the completion of the changeover this information would be analyzed by plotting it on a stacked bar chart.

b)   Plot the results on a stacked bar chart.
(scale it in seconds, do in pencil, one bar for each person)
1. First action at the base of the bar
2. Each subsequent action above it.
3. Color value add action – Green
4. Color non Value add action in RED
5. Estimate the potential improvement and if the work could be done with fewer people.

c)   Review with CO Team
1. Pick an immediate go do improvement
2. Ask operations team to do, a go do after each change over
3. Pick the focus for a full RCO Improvement

Look at the red and:
- Eliminate all unnecessary tasks, motions, movement, steps, or parts that need changing.
- Look for ways to have universal parts.
- Question if something needs changing at all.
- Simplify Make things simple to do.
- Eliminate the need for precise measurement by using jigs, fixtures, stop blocks, gage blocks, detents, etc.

- Improve methods to do tasks by examining the functional requirement and simplify.

Try the Improvements

<u>Establish New Changeover Process</u>

The analysis yields a great number of improvements that require several changeover cycles to validate the improvement. Up to three changeovers should be target during the changeover improvement week.

Schedule and Execute full CO practice for line team

Once all activities of the changeover are documented and displayed on a timing chart more ideas to improve the changeover are identified following a five step approach:

1. **Maximize External effort**
   Observe each Internal Action a second time
   Analyze the Internal Action a second time
   Brain Storm Improvement to allow more external work
   Identify a third improvement to try

2. **Optimize the Internal**
   Observe each Internal Action a second time
   Analyze the Internal Action a second time
   Brain Storm Improvement to speed up internal
   Identify a third improvement to try

3. **Optimize the External**
   Observe each External Action a second time
   Analyze the external Action a second time
   Brain Storm Improvement to reduce the work
   Try Improvement

4. **Eliminate Adjustments**
   Observe each Adjustment
   Brainstorm Improvement
   Apply Poke Yoke
   Implement Improvements

5. **Brainstorm Improvement**
   - No Time-Eliminate time wasters from changeover.
   - No Tools-Eliminate wherever possible the use of a tool.
     Make use of simple, hand-operated devices.
   - No Talent-Make it so simple anyone can do it.

Try the ideas and incorporate the ones that work. Do these trials in rapid but controlled evaluation trials.

Step 3 Tools and Tips
1.	Develop Initial Timing Chart
2.	Clarify Internal and External Actions
3.	Brainstorm Improvements

Chapter 4: Internal Improvement Validation

Step 4: Validate Selected Changeover improvements

True Statistical validation takes several weeks to several months. That level of validation is part of the changeover process but immediate validation comes during the improvement period. The immediate validation is done in the same week when the improvement is achieved. The longer term validation is done as part of the Daily Management System. The short term validation should be based on the measures that will be used to maintain the changeover process.

The proof of improvement rewards those doing the improvement and the business is assured of effort and money well spent.

Organizational Validation Elements

Integration of the changeover progress into the daily management process ensures longer term survival for the improvement.

Training materials, engineering documents, and equipment operation instructions all need to be up dated.

Equipment Validation Elements

Equipment operation and maintenance should be closely scrutinized to ensure the changeover improvement has not cause any production issues. Safety and Quality are the two main elements to validate.

Chapter 4 Tips and Tools

1. Get an analyst to specify the long term validation statistics.
2. Work with the maintenance members to ensure mechanical integrity and ease of maintenance.

3. Do a second quality validation design with the product quality owners.
4. Do a safety evaluation of all actions of the changeover.

Chapter 5: Practice Selected Changeover Improvements

Step 5: Practice Selected Changeover improvements

Practice of the changeover should be scheduled as part of the production process. The only difference between a normal changeover and a practice one is that the practice one has some additional observers watching and timing the changeover. Their observations and feedback is incorporated into keeping the changeover at its peak performance.

Chapter 5 Tips and Tools

1. Use Observation timing sheet.
2. Have experts from other lines be the observers.
3. Plan and schedule practices out into the future.

Chapter 6: Integrate into Daily Management System (DMS)

- Identify Leadership Actions to Maintain RCO Capability

- Identify a sustaining leader changeover partner

- Periodically check changeover time

- Dept Manager Participates 1 time per week

- Operations Manager observes CO 1 time per month

- Evaluate Reapplication of improvement

Chapter 6 Tips and Tools

1. Make the changeover evaluation and time achievement part of the DMS review.

Ron Mueller P.E.

- Integrated Work Systems (IWS) materials author
- Coach to dozens of Manufacturing Directors across the world.
- Certified TPM Coach.
- Tested and proven to enable true breakthrough improvement of Supply Chains.

A proven leader of smart systems implementation across supply, manufacturing and distribution to drive out cost, inefficiencies and to establish synchronized Supply Chains. He utilized the best thinking of Japan's TPM leaders and crafted the necessary related pillars and systems that work in Consumer Products Manufacturing. The results delivered include reduction of Raw and Finished Product Inventories by 40%. Delivered over $100 million is loss reduction through focused systems Workshops across dozens of sites. Developed P&G IWS program materials for external sale. Winner of P&G's Diamond Award for Contribution to Product Supply.

Core Competencies include:
- ✓ Coaching Manufacturing Leadership,
- ✓ Implementation of Integrated Work Systems,
- ✓ Statistical Replenishment design and implementation,
- ✓ Supply Chain Synchronization; author of 3 books in the Stress Free^{TM} series that aid Business and Supply Chain leaders to develop and improve their organization's performance.

Gordon Miller P.E.
- Manufacturing Performance Program
- Development and Delivery Expert.
- Application of Intelligent Manufacturing technology against biggest business challenges with proven business results.

A record as a collaborative and leading-edge thinker, developing programs to deliver cost, productivity and growth enabling manufacturing technology systems deployed via smart standards and empowered teams. As an early developer of PR/OEE measures and improvement programs, has experience with unlocking organization capability for improvement with smart strategies. Led program that developed initial P&G Manufacturing Execution System, leveraged globally across multiple GBUs. Influenced Beauty and Household Care manufacturing systems changes that enabled and leveraged global standardization for rapid footprint growth. Experience that enabled 50% reduction in OEE losses. Experience as a leader of corporate STEM talent strategy can assess and devise approaches to ensure Talent needs for the challenging future are met.

Core Competencies include:
- ✓ Global Productivity Program Design and Management,
- ✓ Advanced Manufacturing Technology Innovation and Strategy Development,
- ✓ Development of Highly Effective Global Teams,
- ✓ Vendor development and management, Organization Capability Development,
- ✓ Talent Strategy

Other books by Ron Mueller

Stress Free™ *Supply Chain Solutions*

Stress Free™ *Manufacturing Solutions*

Stress Free™ *Work Process Solutions*

Stress Free™ *Changeover Solutions*

Stress Free™ *Daily Management Solutions*

Other books by Ron Mueller

41

Ron Mueller

AROUND THE WORLD PUBLISHING, LLC,